W0038224

American Philosophical Society Library
Publication Number 9

Accounts of European Science,
Technology, and Medicine
Written by
American Travelers Abroad,
1735–1860,
in the Collections of the
American Philosophical Society

DARWIN H. STAPLETON

American Philosophical Society Library · Philadelphia · 1985

Copyright © 1985 by American Philosophical Society
for its Library Publications Series,
Number 9.

Cover illustration: Loammi Baldwin, Engineering Diary, 1823
Turning gear for bridge at Brussels /"Bridge at Brussels."

LC: #85-71739
ISBN: 0-87169-350-X

Contents

Foreword

The Andrew W. Mellon Foundation in 1982, seeking to increase the scholarly usefulness of our collections and enrich the intellectual life of the Library, made a grant to the American Philosophical Society establishing a fellowship program in bibliography, research scholarship, and staff development. The major thrust of the program is to provide fellowships for younger scholars who will prepare bibliographic studies on topics well-represented in the Library's collections.

Dr. Darwin H. Stapleton, associate professor of the history of science and technology at Case Western Reserve University, was our first Mellon Fellow in Bibliography in 1983–84. It is hoped that his fine bibliographic study will promote scholarly inquiry in not only the history of science, technology, and medicine but also American Studies in general. Dr. Stapleton was uniquely equipped to undertake this descriptive analysis of the Library's collections, having edited *The Engineering Drawings of Benjamin Henry Latrobe* (1980) and later surveying our holdings in American technology, the report of which appeared in *Technology and Culture* (July 1982).

We would like to acknowledge the generous support of The Andrew W. Mellon Foundation that made possible this work and those that will follow.

April 1985

Edward C. Carter II
Librarian
American Philosophical Society

Preface

The Library of the American Philosophical Society is a marvelous setting for scholarly work. I was therefore excited as well as honored to be asked by the Librarian, Dr. Edward C. Carter II, to serve as a Mellon Bibliographer for 1983–84. The opportunity to study an immense body of sources for the history of American science and technology proved to be all that I had expected.

The entire staff of the Library aided my researches, but I wish to thank some of them in particular. Roy Goodman, Stephen Catlett, and Beth Carroll guided me to numerous sources which I would not have located otherwise and made numerous helpful suggestions, in addition to providing courteous reader service. Discussions with Murphy Smith, Willman Spawn, John Van Horne, Jeffrey Cohen, and Lisa Robinson were uniformly stimulating and supportive. Geri Vickers provided a variety of thoughtful assistance in addition to typing. Dr. Carter's encouragement and enthusiasm for my study were always appreciated.

Darwin H. Stapleton
Philadelphia, June 1984

PART ONE

---

*Preliminary Observations and
Suggestions for Further Study*

I. *The Significance of American Travelers' Accounts.*

From the later colonial period through the period of early industrialization and scientific professionalization, Americans looked to Europe, especially Britain and France, for innovation and inspiration in many fields of endeavor. They could learn much about Europe by talking with recent immigrants, and temporary residents such as ship captains or soldiers, or by reading books and newspapers. But there was no substitute for traveling abroad and keeping a careful record of what was seen, heard, and done. And on return it was appropriate to share that knowledge with one's peers.

The need for direct acquaintance with the state of European knowledge was nowhere more acute than in the fields of science, medicine, and technology. The eighteenth and nineteenth centuries' rapid advances in these fields outstripped the abilities of conscientious American correspondents or energetic booksellers. Moreover, there were certain aspects of science, medicine, and technology which had to be experienced in person. In science the comraderie of fellow specialists was seldom available on this side of the Atlantic until well into the nineteenth century; scientifically-active Americans cherished the personal respect they found among those Europeans who knew their work. But there was also a more practical value in direct contact with European science: the use of laboratory apparatus and the conduct of elegant classroom demonstrations required hands-on experience. Numerous Americans with fine scientific training were enthralled by the scientific instrument shops and lecture halls of London and Paris.[1]

In medicine and surgery Americans found European experience invaluable. Many who were already well-trained in medical doctrine and materia medica found that several months to a year in a large London or Parisian hospital provided them with opportunities to observe diseases or take part in surgical operations which would have required years in smaller American population centers.

Personal contact was of even greater importance in various areas of technology. As Eugene S. Ferguson pointed out in a classic article, there is

1. I think particularly of Alexander Dallas Bache, Joseph Henry, and Alfred Perkins Rockwell, whose accounts are in the following bibliography.

much technical knowledge which cannot be reduced to the spoken word and needs to be learned by direct observation or by on-the-job training.[2] Thus technically-inclined Americans who wanted to be truly up-to-date went abroad and actually observed (or even took part in) industrial processes and technologies. Frequently they kept journals or diaries which supplemented verbal summaries with pictorial or diagrammatic representations.

The flow of American scientific, technical, and medical visitors prior to 1860 (the terminal date for this study) was, of course, subsumed within a far greater number of Americans abroad for other purposes, including diplomacy, business, pleasure, and other studies. Often the travelers mixed observations of science, technology or medicine with one or more other interests. There are several bibliographies or studies which provide access to American travel accounts.[3]

My focus has been on those Americans who made substantial written comments (journals, diaries, notebooks, letters or published accounts) on science, technology, or medicine, usually on the basis of their training or experience in those fields before they left America. These informed travelers' accounts are valuable to historians for at least three reasons: (1) they often provide detailed information about European personalities, sites, or techniques; (2) they are closely associated with the transfer of technology, scientific ideas and methodology, and medical and surgical techniques from Europe to America (and occasionally vice-versa); and (3) by the travelers' comparative and other statements they provide an index of the state of American science, technology, and medicine.

Previous utilization of travelers' accounts for these purposes has been less than one might expect. A sufficient reason may be that the significance of foreign travel in an individual's training and education has been inadequately understood. In preparing this study I have noticed that European travel has

2. Eugene S. Ferguson, "The Mind's Eye: Nonverbal Thought in Technology," *Science* 197 (26 August 1977): 827–36.

3. William L. Sachse, *The Colonial American in Britain* (Madison, Wis.: University of Wisconsin Press, 1956); Paul R. Baker, *The Fortunate Pilgrims: Americans in Italy, 1800–1860* (Cambridge, Mass.: Harvard University Press, 1964); Bruce Sinclair, "Americans Abroad: Science and Cultural Nationalism in the Early Nineteenth Century," *The Sciences in the American Context: New Perspectives* (1979) and Russell M. Jones, "American Doctors in Paris, 1820–1861," *Journal of the History of Medicine* 25 (April 1970): 143–57 (including a 5-page bibliography of visitors' accounts), are the most recent studies I have seen. Older works include: Anna M. Babey, *Americans in Russia, 1776–1917* (New York: The Comet Press, 1938); Whitfield J. Bell, Jr., "Philadelphia Medical Students in Europe, 1750–1800," *Pennsylvania Magazine of History and Biography* 67 (January 1943): 1–29; David Sanders Clark, Comp., "American Travelers and Observers in the British Isles, 1850–1875, A Bibliography" (Cleveland, Oh.: Typescript, 1940); Carrie Evangeline Farnham, *American Travellers in Spain: the Spanish Inns, 1776–1876* (New York: Columbia University Press, 1921); Howard R. Marraro, "American Travellers in Rome," *Catholic Historical Review* 19 (1944): 470–509; Robert Belmain Mowat, *Americans in England* (Boston: Houghton Mifflin Co., 1935); Robert C. L. Scott, "American Travellers in France, 1830–1860" (Ph.D. dissertation, Yale University, 1940).

often been omitted from biographical sketches in standard reference works, or if mentioned, little is made of it. Authors seem to regard European travel as more akin to leisure than to education—yet the accounts themselves indicate dedicated (occasionally frenzied) gathering of specimens, visits to industrial works, and meetings with eminent Europeans. I think it is worth quoting the dean of mid-19th century American science, Benjamin Silliman, Sr., who looked back on his European visit of 1805–06 with the perspective of 45 years and commented: "The year I passed in Great Britain in my youth [was] the most instructive year of my life."[4] For many Americans the fund of observation and experience acquired had profound effects on their careers and interests.

Another reason for many historians' relative indifference to foreign travel is certainly the national (even nationalist) framework within which much history is written. To admit the significance of foreign travel implies recognition as well of the international context of knowledge and action. Allowing for the international flow of technical ideas, for instance, calls into question the "invention" of numerous devices by recently-returned Americans. To use a hoary example, the textbook stories of Robert Fulton's first steamboat normally fail to point out that his technical ideas and his practical knowledge were developed during his twenty years abroad.

In general, then, I expect that further study of Americans' accounts of European science and technology will deepen historians' understanding of the generation and flow of ideas and techniques in the North Atlantic community in the century or so prior to 1860. Since that time period coincides with a period of rapid industrialization, the rise of scientific professionalization, and the transformation of medicine and surgical practice, the fruits of such study would be immense.

II. The Nature of the Accounts

Travelers' accounts were kept for varying purposes and in varying forms, but it is possible to categorize them roughly in order to grasp which kind of sources they are.[5] Fundamentally, the accounts reflect the reasons the writers traveled. Some, for example, were commissioned to go abroad to gather information, and they were often expected to submit formal reports on their findings. One of the earliest instances of government support was in 1815

4. Benjamin Silliman, *A Visit to Europe in 1851.* 2 vols. (New York: G. P. Putnam and Company, 1854), 1: 78.
5. In what follows unfootnoted sources are those which are included in the bibliography.

5

when the Department of War sent Major Sylvanus Thayer abroad to study advanced engineering. His instructions read:

. . . you will proceed to the Continent and prosecute those enquiries and examinations calculated for your improvement in the military art. The military schools, workshops, and arsenals, the canals and harbors, the fortifications, expecially those for maritime defense will claim your particular attention.[6]

Thayer's European knowledge was deemed so valuable that he was on his return appointed superintendent of the Military Academy at West Point. The Department of War subsequently sent a number of other officers abroad, but whether it had an established policy of doing so I do not know.[7]

Although Benjamin Silliman went abroad under Yale's auspices in 1805–06, it was a decade after Thayer's departure until private organizations regularly gave commissions. Beginning with William Strickland's 1825 trip to England at the behest of a group of Pennsylvania internal improvement enthusiasts, there was a regular succession of trips, including those of the Baltimore & Ohio Railroad engineers, Alexander Dallas Bache, Samuel V. Merrick, Wilbur Fiske, and Benjamin Franklin Peale. The accounts of commissioned visitors were often published and at their best represent landmark assessments of European technology and science. For example, Ellis Sylvester Chesbrough's report on European sewerage systems of 1856–57 (for the Chicago Board of Sewerage Commissioners) was the first comprehensive statement of sanitary engineering practice available to Americans.[8]

A larger body of material is that produced by Americans who were abroad for formal study. The bulk of this material, and certainly the most thoroughly studied, is that of physicians and surgeons. Whitfield J. Bell, Jr., and Russell Jones have pursued American physicians in Britain and France, the latter concluding that there were at least 616 Americans studying in Paris from 1820 to 1861.[9] Other compilations of American students yield smaller numbers, but since we have studies for only a few schools and since students attending public lectures (such as Moncure Robinson) or paying for tutorials

6. Department of War to Sylvanus Thayer, 20 April 1815, reel 2, film 417, National Archives, Washington, D.C.

7. Others include Dennis Hart Mahan, George Wurtz Hughes (see no. 17 in bibliography), Richard Delafield, Alfred Mordecai, and George B. McClellan.

8. Louis P. Cain, "Raising and Watering a City: Ellis Sylvester Chesbrough and Chicago's First Sanitation System," *Technology and Culture* 13 (July 1972): 363.

9. Jones, "American Doctors in Paris," p. 150; Russell M. Jones, "American Doctors and the Parisian Medical World, 1830–1840," *Bulletin of the History of Medicine* 47 (1973): 40–65, 177–204. The latter article revises Jones's estimate of 178 Americans studying medicine in Paris in the 1830s upward to 222; presumably his estimates for the other decades of 1820–1861 have risen also. Whitfield Bell's publications may be most conveniently located in the catalogue of the American Philosophical Society Library.

are not included, estimates of their numbers are likely to be low.[10] I think that diligent research on the French engineering schools after 1815, for example, would turn up a surprising number of Americans.

In any case, students were necessarily literate, normally kept lecture notes, and often wrote home to inform those who were paying tuition and board how their education was progressing. Ensconced in some of the best educational institutions in the world American students often described famous figures of science, engineering, and medicine, and their class notes may sometimes record the formulation of their mentors' scientific and technical ideas before they were published.

The bulk of scientifically-and-technically-inclined Americans abroad, however, were neither commissioned experts nor students, but were engaged in pursuing personal studies and goals. John Griscom, George Escol Sellers, Joseph Henry, and Maria Mitchell, for example, went abroad armed with letters of introduction and mental lists of sights to see. Traveling as their resources and linguistic skills allowed, these Americans stayed abroad several years (William Maclure) or for a few months (Sellers).

If these three types of travel accounts tend to share any common element, it is in the sights they record. As strangers in strange lands they tended to seek out the places already known to them from history or previous travelers' accounts, in part no doubt because they knew they would be asked about them on their return. Some sights seemed to be perennial (Vesuvius, Stratford-on-Avon, French cathedrals), but some clearly changed with the times. The Duke of Bridgewater's canal was an attraction of the 1770s and 1780s, but the Liverpool and Manchester Railroad was a mecca in the latter 1820s and the 1830s. Likewise, Herschel's great telescopes at Slough made it a stopping place of the late eighteenth and early nineteenth century, but later the Greenwich Observatory became an object of admiration.

As residents of a country with relatively few cultural institutions and few persons with the wealth to devote to amateur science, Americans were also fascinated by European museums and collections. Some, like the wax anatomical models in Florence and Vienna, had no equal in America, and excited almost uniform admiration. Collections of scientific instruments and models of machinery also brought enthusiastic responses, particularly the Conservatoire des Arts et Metiérs. John Griscom thought it the best arranged and presented exhibit of its type in the world. Zachariah Allen thought it was

---

10. R. W. Innes, *English-Speaking Students at the University of Leyden* (Edinburgh and London: Oliver and Boyd, 1932) lists 21 students from the North Atlantic coastal colonies and (later) the United States, 1729–1795. Margaret W. Rossiter, *The Emergence of Agricultural Science: Justus Liebig and the Americans, 1840–1880* (New Haven and London: Yale University Press, 1975), pp. 186–192, lists 33 students studying under Liebig at Giessen and Munich, 1841–1860.

equaled by the Patent Office collection in Washington, but believed that both were excellent sources of mechanical instruction.

Over the years thousands of Americans toured industrial and engineering works throughout Europe, but especially in Britain. Military depots and government works were obvious choices. Many saw the Arsenal at Venice (now long past its years of glory but still functioning) no doubt because it was described by Dante and Galileo. The porcelain works at Sèvres and the Gobelin tapestry workshop were favorite stops in France. The Portsmouth and Plymouth naval yards attracted attention in England, but after the Revolution subterfuge was required, since only citizens of the Empire could be admitted to British military and naval sites.

The greatest technological attractions for Americans, however, were the wonders of the British industrial revolution. Numerous Americans, such as George Escol Sellers, frankly made the observation of special machinery and processes the object of their visits. Others were omnivorous and, like Joshua Gilpin, took in everything they were able to see. Most were impressed by the evident superiority of British technology: Thomas Jefferson's visit of 1786 moved him to remark that "the mechanical arts in London are carried to a wonderful perfection."[11] Fewer American commentators took an interest in agriculture, though Jefferson was an exception, as was Robert R. Livingston. Probably the bulk of those writing accounts were urbanites, and only the landed gentry (a small class in America) were likely to find the agrarian arts instructive.

For the scientifically-, technically-, and medically-inclined, meeting Europeans of similar inclinations was of equal importance to sightseeing. Most Americans came with some letters of introduction, often written by their predecessors. Joseph C. Naurede drew upon his acquaintance of 18 years earlier when he wrote to Baron Dominique-Jean Larrey to recommend to him an American surgeon visiting Paris in 1840.[12] Those who needed letters could usually obtain them from American ambassadors or consuls, or from Americans permanently abroad, such as Benjamin West, Charles Leslie, Samuel I. Fisher, or David Bailie Warden.

In any case, Americans were generally well received by their European contemporaries. Some Americans seem to have enjoyed meeting as many of the great and near-great as possible: whether out of self-aggrandizement or a sense of professional obligation it is not easy to say. Often the encounters were brief, but even brief ones could leave lasting impressions, as Maria Mitchell's notes on a half-hour encounter with Humboldt demonstrate.

11. Thomas Jefferson, *The Papers of Thomas Jefferson.* Julian P. Boyd and Charles T. Cullen, eds. 21 vols. to date (Princeton, N. J.: Princeton University Press, 1950–), 9: 445.
12. Joseph G. Naurede, 2 December 1840, to Larrey, Larrey Letters, microfilm, American Philosophical Society Library.

A surprising number of Americans attended professional meetings abroad. A few, such as Franklin or Warden, were long resident in European capitals and regularly attended meetings of the Royal Society and the Institute de France. American medical students often joined local medical societies (some of which were specifically for students), and they even founded their own American Medical Society in mid-nineteenth-century Paris. Occasionally traveling Americans of some distinction were elected corresponding or foreign members of learned societies which they visited: John James Audubon seems to have collected memberships during his travels in Britain as easily as he collected bird specimens in America.

The annual meetings of the British Association for the Advancement of Science (beginning in 1831) were a particular attraction for Americans, undoubtedly because for some time they had no similar group of their own. Several Americans made BAAS meetings a prime object of their visit abroad, and I think it likely that no meeting of the 1830s lacked a contingent of Americans. The impact of seeing such an immense body of the erudite was summarized by Wilbur Fisk after attending the 1836 BAAS meeting at Bristol: "Never did I before so fully realize what was meant by 'the feast of reason and the flow of soul' as during my attendance upon these meetings of the British Association. I never expect again to enjoy the like . . ."[13] Fisk's tribute to the BAAS was echoed, though less emotionally, by travelers such as Joseph Henry, Alexander Dallas Bache, and William Gibson.

As a final note on the nature of the sources, let us consider for whom the accounts were written. Without doubt, most accounts were kept for personal reasons, primarily for personal enjoyment and reference. Closely related are groups of letters written home or journals kept for a spouse or intimate friend. The most systematically or conscientiously kept accounts were probably the more treasured and consequently have tended to be preserved. A few, certainly, were kept with a view toward publication, since travel accounts of all sorts were very popular in the period under study here, and presses seem to have been kept busy printing or reprinting them.[14]

The published accounts, indeed, partake of the tastes of the times by devoting considerable space to observations of antiquities (particularly in Italy) or of exotic places and cultures (especially Turkey and Egypt, though they are not European). There was clearly a regional bias at work: Valentine Mott opened his book by telling readers that in writing about northern Europe he would "embrace almost exclusively matters seldom dwelt upon by

13. Wilbur Fisk, *Travels in Europe; viz., in England, Ireland, Scotland, France, Italy, Switzerland, Germany, and the Netherlands.* 4th ed. (New York: Harper & Brothers, 1838), p. 612.

14. In addition to the sources in note 3, a useful approach to travel literature is Harold F. Smith's *American Travellers Abroad: A Bibliography of Accounts Published Before 1900* (Carbondale-Edwardsville, Ill.: The Library, Southern Illinois University, 1969).

tourists, and related to medical science" but "as we advance into the more ancient countries of Southern Europe and the East, the degraded condition of medicine there . . . furnish again occasion to revert to the prouder epochs of their history in bygone ages."[15] Mott thus made assumptions about what his audience wanted to hear (and what would make his book sell) which profoundly affect his book's content and its value as a historical source. The writers of travel accounts, like historians, had distinct points of view which should be assessed before the accounts are put to use as historical evidence.

III. *Questions Raised and Questions Answered*

The study of American accounts of European science, technology, and medicine up to 1860 can assist historians in understanding a number of important historical developments. Most important is the international flow of knowledge and innovation. The growth and development of the American medical and surgical professions were perhaps most heavily influenced by European precedents, since entrance into them was largely controlled by medical schools dominated by European-trained professors.

Science was affected less consistently, since it was possible for autodidacts like John Bartram to emerge as significant figures in the eighteenth century, and for major figures like Joseph Henry to acquire international stature without European training. Yet the American Philosophical Society, a major focus of scientific activity throughout the period, thrived on the international contacts and international travels of its members. And numerous institutions, such as astronomical observatories and the American Association for the Advancement of Science (established in 1848) were consciously modeled after European precedents.

The importance of transfers of European technology to the United States is well-known and can scarcely be overestimated. Without question historians can discern early in colonial times the signs of distinct American technologies in agriculture and waterpowered milling. Yet it is symbolic that in northern Delaware where the American genius Oliver Evans developed the automatic flour mill, there was established contemporaneously a gunpowder works of advanced form which Eleuthére Irénée du Pont regarded as "a colony" of the French Administration of Powder and Saltpeter.[16]

The importance of the transfer of technology has recently been cast in a

15. Valentine Mott, *Travels in Europe and the East* (New York: Harper & Brothers, 1842), pp. vi–vii.

16. Darwin H. Stapleton, "The Transfer of Technology to the United States in the Nineteenth Century" (Ph.D. dissertation, University of Delaware, 1975), p. 84.

new form by Anthony Wallace. Trying to come to grips with the flow of textile technology ideas between Europe and America, Wallace asserted that there was in the first half of the nineteenth century an "international fraternity of mechanicians" in the English-speaking world. A group of a few hundred skilled machine builders, he concluded, knew each other by reputation, writings, and to a large degree by personal contact.[17] Certainly the accounts written by Zachariah Allen and George Escol Sellers support Wallace's argument, though few mechanics were as verbal as they were.

In American science, technology, and medicine, then, international contacts were extremely important. Travelers' accounts are significant to a large degree because they provide information on how the international contacts were made and because they sometimes indicate their effects. We know, for example, that during his visit to Britain in 1825 William Strickland established a long-term friendship with civil engineer Jesse Hartley, and that Hartley was subsequently a gracious host to other American engineers, including Horatio Allen, the B&O engineers, George Escol Sellers, and Samuel V. Merrick.[18] Similarly the eighteenth-century Quaker physician in London, Dr. John Fothergill, routinely provided advice and introductions to American students, while the nineteenth-century Belgian astronomer L. A. J. Quetelet was always ready to welcome visiting Americans.

All told, my impression is that surviving travel accounts, as well as our existing knowledge of diffusion of ideas, are just the visible tenth of an iceberg. First, we probably do not have (and may never have) anything close to an accurate count of visitors.[19] Secondly, we have not fully explored the means by which new scientific ideas were spread, new technologies were transferred, or advanced medical and surgical methodologies were taught, although the very accounts which are the subject of this essay were one means. Research in both of these areas will do much to promote our understanding of the growth of American science, technology, and medicine to 1860.

Moving from this larger issue, let us turn to four specific topics on which travelers' accounts can shed considerable light. One that I have been particularly sensitive to is the extent to which Americans were prevented from learning crucial industrial secrets abroad. If restrictions were significant, then the value of travelers' accounts (and the transfer of technology) would be limited. The question of secrecy is raised by comments such as that by Samuel

17. Anthony F. C. Wallace, *Rockdale* (New York: Knopf, 1978), pp. 211–19.

18. "Diary of Horatio Allen: 1828 (England)," *Bulletin of the Railway & Locomotive Historical Society* 89 (November 1953): 100–04.

19. Note Eugene S. Ferguson's comment (first published in 1962) that "The stream of travelers going to Europe to obtain mechanical and engineering information was important, but its magnitude is not known with any precision. I suspect that it was perhaps five times as great as the best-informed scholars today would estimate it to be." Edwin T. Layton, Jr., ed., *Technology and Social Change in America* (New York: Harper & Row, 1973), pp. 23–24.

Curwen (1777) that "it is with difficulty that one is admitted to see their works [at Manchester], and in their many cases it is impracticable, express prohibitions being given by their masters." Benjamin Franklin Peale, abroad in 1834, found that private coining and rare metal businesses in France and England were "exceedingly close."[20]

But for every case like these there were others (I think many others) in which Americans had the opportunity to inspect thoroughly industrial processes. Even supposedly uncooperative machine manufacturers became cooperative in the right circumstances (as George Escol Sellers found with Bryan Donkin), and civil engineers and textile mill operators seem to have been uniformly responsive to requests for information. My impression is that well-intentioned Americans, particularly those who could be identified as businessmen (such as Joshua Gilpin), or as fellow members of "the international fraternity of mechanicians" (such as Zachariah Allen) seldom had difficulty obtaining the information they sought. Occasional obstinancy was often overcome by more-or-less deliberate industrial espionage.

Another issue which the use of travelers' accounts might address is the role of women in disseminating scientific and medical ideas to Americans. John Morgan's journal, one of the earliest accounts by an American physician in Italy, describes a meeting with Dr. Laura Maria Catherina Bossi, a distinguished physical scientist at the University of Bologna. (Nearly a century later Silliman visited Bologna and noted that university's tradition of women professors.) Encounters with women of some reputation and skill in science or medicine occur with some regularity thereafter. Often they took place in the context of a learned household, such as the Herschels or the Murchesons, but sometimes Americans sought out independent interviews, such as Isaac Lea's visit with "Mrs. Marie to see her beautiful collection of shells," or Maria Mitchell's fascinating meeting with Mary Somerville. If (the mostly male) travelers' accounts sometimes carry a note of surprise or even a sexist remark on encountering their intellectual equals in feminine form, they nevertheless indicate that they were required to come to grips with the international *brotherhood* of science in a way which was perhaps not so common on this side of the Atlantic.

As this point suggests, travelers' accounts are excellent sources for comparative studies. Clearly one can find numerous comparative statements which measure the level of American education, institutions, or achievements by European standards. Sometimes these assessments seem motivated as much by patriotic pride as by realism, but when both Zachariah Allen (1825)

20. Samuel Curwen, *Journal and Letters of the late Samuel Curwen, Judge of the Admiralty,* etc., *An American Refugee in England from 1775 to 1784* . . . George Atkinson Ward, ed. (New York: C. S. Francis and Co., 1842), p. 136; Benjamin Franklin Peale, 13 October 1834, to Moore, IX A/4B13, in *The Collected Papers of Charles Willson Peale and His Family.* Lillian B. Miller, ed. (Millwood, N.Y.: Kraus Microform, 1980).

and George Escol Sellers (1832) found that American machines and machine tools already exhibited distinctive and sometimes superior qualities when compared to British examples, it seems safe to go on to determine why there was a difference. I also see no reason to take exception to a series of remarks by American physicians abroad that from the late colonial period onward the theory of medicine was as well taught in leading American colleges as in London, Edinburgh, or Paris. Those remarks are in themselves powerful reminders of the success of and continual renewal of transatlantic flows of knowledge. Should there be any questions about the relative standing of European and American science, technology or medicine, travelers' accounts written by informed Americans should go a long way toward providing answers.

There is a related issue that can be addressed as well. What were the motivations for Americans who went abroad, in addition to acquiring new knowledge? Clearly for many Americans a trip abroad was a symbol of status. In medicine or surgery, European experience (whether it improved one's methodology or technique) was regarded as a means of attracting a more extensive (or more elite) clientele. In that case it took money to make money: John Morgan reported that his three years of study, 1763–65, cost £1500.

For others some time abroad provided the opportunity for leisure or, as Thomas Jefferson feared, dissipation. Moncure Robinson remarked on how easy it was for students in Paris to enjoy the cafes more than the classroom, and numerous letters of recommendation in the Warden Papers are frank about a visitor's intent to combine professional improvement with pleasure.

Finally it should be noted that numerous institutions in Paris provided free public lectures which were of very high quality. Although the costs of transatlantic travel and residence in Paris were certainly not low, the saving of tuition was frequently touted as a major reason for studying abroad.

In sum, American accounts of European science, technology, and medicine may provide new resources for examining issues and questions of importance to scholars. The following annotated bibliography of sources provides a substantial sample of such accounts because they were accumulated by an institution whose members cultivated international communication. I believe that scholarly understanding of sources like these will aid in continuing the development of international communication today, in a world that sorely needs understanding and cooperation.

PART TWO

---

*An Annotated Bibliography of
Printed and Manuscript Holdings at
the American Philosophical Society Library*

KEY TO MARKS AND ABBREVIATIONS

*   Member of the American Philosophical Society

Appleton's   John Grant Wilson and John Fiske, eds. Appleton's *Cyclopedia of American Biography*. 6 vols. (New York: D. Appleton and Co., 1888–1889).

APS   American Philosophical Society

BDAS   Clark A. Elliott, *Biographical Dictionary of American Scientists* (Westport, Conn.: Greenwood Press, 1979).

DAB   Allen Johnson and Dumas Malone, eds., *Dictionary of American Biography*. 20 vols. (New York: Scribners', 1928–1936).

DSB   Charles C. Gillispie, ed., *Dictionary of Scientific Biography*. 16 vols. (New York: Scribners', 1970–1980).

BIBLIOGRAPHER'S NOTE

The annotated bibliography results from several months' search of the holdings of the American Philosophical Society's Library, and a thorough examination of each source's contents. I believe that there are additional significant travelers' accounts at the APS yet unlocated because one cannot expect card catalogues or manuscript finding aids to list all items, or to list them in terms which make all contents obvious. Several items on this list were located by clues, scholarly intuition, or hints from the Library's staff rather than clear references.

At the end of each bibliographic entry I have noted (if they are not apparent from the title of the document) the date of the account, whether it is accompanied by significant illustrations, and where the travel took place.

The terminal dates for this study derive from the collection strengths of the Library and from my own interests. I certainly believe that similar travelers' accounts up to the recent past (I think James Watson's *The Double Helix* fits the genre) are equally worthy of study.

Printed Works

1. Allen, Zachariah, *The Practical Tourist, or Sketches of the State of the Useful Arts, and of Society, Scenery &c &c in Great-Britain, France and Holland.* 2 vols. Providence, R.I.: A. S. Beckwith, 1832. Some illustrations. England, Wales, Scotland, Ireland, France, Netherlands.

Allen (1794–1882, *DAB*) was a Rhode Island textile manufacturer and inventor of mechanical devices. He took his trip abroad in 1825, though his account includes some information of later date. While Allen's interests obviously centered on textiles, he visited a wide variety of factories, engineering works, museums, and institutions, including hospitals and anatomical collections. He did not hesitate to make pronouncements about the quality of products or to compare European efforts to American, making his book particularly valuable for a consideration of early American industrialization.

For example, on visiting a large English flour mill at Leeds he noted "I saw here none of the labor-saving contrivances for receiving the grain and elevating it by machinery to the grain lofts, without manual labor, in little leather buckets fixed on an endless revolving band; nor the perfect arrangements for removing hot flour as fast as it falls from the stones, and for stirring and cooling it previously to its being packed." (1:205). In France he paid more attention to agriculture than in Britain, but he visited and admired the mechanical models at the Conservatoire des Arts et Metiérs, and took the time to inform a woolen manufacturer at Louviers about American innovations. On a brief visit to the Netherlands he took an interest in canals, windmills, the National Manufactures exhibition at Haarlem, and Tyler's Museum. This work is a real compendium of European technology, having many brief but effective descriptions. Americans of the time thought well of it (see review in *American Journal of Science* 23 [1833]: 213–25), and historians of the period have found it a useful reference.

2.   *Bache, Alexander Dallas, *Report on Education in Europe to the Trustees of the Girard College for Orphans.* Philadelphia: Lydia R. Bailey, 1839. England, Germany, Scotland, Netherlands, Austria, Switzerland, France, Belgium, Italy.

Bache (1806–67, *DAB*) was a graduate of West Point and a professor of natural philosophy and chemistry at the University of Pennsylvania before being chosen first president of Girard College in 1836. (At his death in 1831 Philadelphia merchant Stephen Girard left a bequest to establish a school for orphans.) The College trustees soon thereafter voted to send Bache to Europe to study orphan and educational institutions so that on his return to America he could establish the most enlightened curriculum and regulations possible. Bache spent two years abroad (September 1836–October 1838) visiting a wide variety of institutions and indulging his catholic interests in science and technology (see no. 49 below).

For science and technology the most interesting material in this massive work (666 pp.) is Chapter XIII, "Superior Schools," which describes the École Polytechnique at some length, as well as the School of Arts and Manufactures at Paris, the Boarding Institute of Arts at Charonne, the Schools of Arts in Prussia, the Institute of Arts at Berlin, the Polytechnic Institute of Vienna, the School of Mines of Saxony at Freiberg, the Institute of Agriculture and Forestry at Hohenheim, and the Naval School of Austria. Much of the material presented are translations and transcriptions of school documents, but Bache usually commented on who hosted him during his visit, described what he saw, and offered opinions on the curricula and institutional settings.

3. Baldwin, George Rumford, letter of 8 June 1834 to Loammi Baldwin, in Loammi Baldwin, *Report on the Subject of Introducing Pure Water into the City of Boston.* Boston: John H. Eastburn, 1834. Pp. 16–18. Scotland.

    George R. Baldwin, brother of the better-known Loammi Baldwin, related in this letter his knowledge of the Edinburgh waterworks (constructed by James Jardine) which he examined in 1832. Elsewhere in this booklet Loammi mentioned but did not describe his earlier examinations of the Edinburgh and Paris waterworks, presumably made during his trip abroad in 1823–24 (see no. 50 below).

4. *Booth, James Curtis. Wyndam D. Miles, "With James Curtis Booth in Europe, 1834," *Chymia* 11 (1966): 139–49. Austria.

    This article includes substantial extracts from Booth's journal of his visit to Vienna during his German studies of the 1830s. Booth (1810–1888, *DAB*) showed particular interest in industrial works, including a hat works, paper mill, tannery, almond oil press, a textile printing factory, and a porcelain works. He also visited the Polytechnic Institute and several museums.

5. Colburn, Zerah, [Cost, Working, and Construction of English Railways], *Journal of the Franklin Institute,* 3rd ser., 35 (April 1858): 285–87.

    Colburn (1832–70, *Appleton's*) was a publisher and mechanical engineer. He had visited England and France in 1855 and 1856, and during and after his trip published in the *Railroad Advocate* (of which he was editor) accounts of iron and machine works there. In 1857 "he again visited Europe at the request of several railroad presidents" (*Appleton's*) and in 1858 published a work comparing American and European railroad practices, titled *The Permanent Way,* coauthored by his traveling companion, the engineer Alexander Lyman Holley. Colburn attended a meeting of the Franklin Institute on 18 March 1858 and presented a series of statistical and qualitative comparisons between American and European railroads, and a description of smokeless coal-burning locomotives used for passenger trains in England. Colburn was apparently traveling to major railroad centers selling subscriptions for his book.

6. Curwen, Samuel, *Journal and Letters of the late Samuel Curwen, Judge of the Admiralty, etc., An American Refugee in England from 1775 to 1784.* George Atkinson Ward, ed. New York: C. S. Francis and C., 1842. England.

Curwen (1715–1802, *DAB*) was a Massachusetts native and graduate of Harvard. He had been abroad in the late 1730s and became a merchant in Salem on his return. At the onset of the American Revolution he was exiled for his Tory sympathies, and spent the duration of the Revolutionary War living and traveling in England. Several times he visited British industrial works and briefly described them, although he found at Manchester that "it is with difficulty one is admitted to see their works, and in many cases it is impracticable, express prohibitions being given by their masters." (7 June 1777) He also attended some public scientific lectures and at least one meeting of the Royal Society (20 March 1783).

7. "Description of the Chain Bridge over the Straits of Menai, North Wales," *Journal of the Franklin Institute* 3 (1827): 138–39.

This note is reprinted from a New York newspaper and is introduced by this statement: "The following description of the stupendous chain bridge in North Wales, is furnished by a friend, who has lately received it in a letter from a gentleman now travelling in England." The publication of such accounts was common in the 1820s when internal improvements agitation reached a peak in the United States. Thomas Telford's chain bridge over the Menai Straits, completed in 1825, quickly became an attraction for technically-inclined tourists.

8. Fisk, Wilbur, *Travels in Europe: viz., in England, Ireland, Scotland, France, Italy, Switzerland, Germany, and the Netherlands.* 4th ed. New York: Harper & Brothers, 1838. Plates of a Swiss suspension bridge, and of the Thames Tunnel.

Fisk (1792–1839, *DAB*) was a New England Methodist minister who became the first president of Wesleyan University of Middletown, Connecticut in 1830. In 1835 he went abroad for several reasons: to restore his declining health, to strengthen relations between American and English Methodists, to study educational institutions, and to purchase scientific apparatus for Wesleyan. His book is written as a series of dated letters followed by undated chapters. It is disappointing for the historian of science and technology to find this work written more as a travelogue than as a description of how he fulfilled the purposes of his visit. His most substantial contributions are his consideration of French educational institutions and of a British Association for the Advancement of Science meeting in 1836. He made only a brief comment on English and French instrument-makers, and described his visits to several manufacturing establishments in England and France.

9. *Franklin, Benjamin, *The Papers of Benjamin Franklin*. Leonard Labaree and William B. Willcox, eds. 23 volumes to date. New Haven and London: Yale University Press, 1959–   .

From 1757 to 1785 Franklin (1706–90, *DAB*) was almost constantly abroad, at first in England (to 1775), and then in France (from 1777). There is little in the *Papers* which can be called a traveler's account, though he loved to travel, making tours through Holland, Germany, and France, as well as part of England. Instead Franklin's *Papers* stand as a reference point for American knowledge of European science and technology. When he received a letter from Philadelphian Samuel Rhoads in 1771 asking about canal construction, for example, Franklin was able to provide documents and a clear statement from personal experience about the preferable technology. As an active member of the Royal Society in London he was a conduit for scientific information flowing to and from America. Thus, any researcher concerned about American knowledge of European science and technology during the Franklin era would be wise to consult his papers.

10. *Gibbs, [Oliver] Wolcott, "Great Exhibition in London," *American Journal of Science,* 2nd ser., 12 (November 1851): 440–42.

Gibbs (1822–1908, *DAB*) had a medical education at New York, then studied in Germany and France, 1845–48. Thereafter he taught science at the Free Academy in New York, and, from 1863–87, at Harvard. On visiting the Crystal Palace Exhibition at London he wrote this report on scientific instruments displayed by the British, French, Dutch, German, and Swiss manufacturers. He noted that the United States Coast survey exhibited two balances by Joseph Saxton "at least equal in point of workmanship to any which the writer has seen abroad." One of those instruments may be that in Library Hall of the American Philosophical Society.

11. *Gibson, William, *Rambles in Europe in 1839*. Philadelphia: Lea and Blanchard, 1841.

Gibson (1788–1868, *DAB*) was an American surgeon who had studied at Edinburgh and London 1806–09, then taught at the University of Maryland and the University of Pennsylvania. He was recognized for his innovations in surgical technique such that by the time he went abroad again in 1839 he was (by his own account) greeted enthusiastically by surgeons in England, Scotland, Ireland, and France. As a result this book reads like a *Who's Who* of contemporary European surgeons and physicians, with frequent descriptions of individuals and their work, often accompanied by anecdotes. There are particularly detailed accounts of the French surgeon Velpeau, and the English surgeon Sir Charles Bell. Gibson described numerous hospitals, medical schools, and museums, and provided extensive descriptions of his attendance at meetings of the Provincial Medical and Surgical Association (July 1839), and the British Association for the Advancement of Science (August 1839). Altogether this book is a stunning record of the international medical community.

12.   *Gilpin, Joshua, extracts of journals of 1795–1801. Harold B. Hancock and Norman B. Wilkinson, "Joshua Gilpin: An American Manufacturer in England and Wales, 1795–1801—Part 1," Newcomen Society *Transactions* 32 (1959–60): 15–28; "Part 2," Newcomen Society *Transactions* 33 (1960–61): 57–66. England and Wales. Plate of Wilkinson's boring mill in vol. 32.

Gilpin (1765–1840, *Appleton's*) was a Philadelphia merchant with a deep interest in industrial technology. During this sojourn he kept detailed journals of his observations: the original journals are at the William Penn Archives (Harrisburg, Pa.), and microfilm copies are at the Science Museum and Friends Library in London. The extracts printed here run to about a page each and cover potteries, roads and canals, coal mining, iron-making, copper manufacture, shot manufacture, pin-making, the Birmingham mint, paper mills, salt mining, glass manufacture, shipyards, and textiles. Gilpin certainly put his newly-acquired knowledge to use. As a result of this and a later trip to England in 1811–1814, he brought chlorine bleaching and machine-made paper into his family business, and he became a tireless promoter of the Chesapeake and Delaware Canal. In 1804 the engineer Benjamin Henry Latrobe said that of all the members of the canal's board of managers, Gilpin was the only one with the requisite knowledge of canal technology.

13.   *Griscom, John, *A Year in Europe, Comprising a Journal of Observations in England, Scotland, Ireland, France, Switzerland, the North of Italy, and Holland, In 1818 and 1819*. 2nd ed. 2 vols. New York: Abraham Paul, 1824.

Written as a series of 42 dated letters, this work is distinguished by the author's contacts with notable European chemists. Griscom (1774–1852, *DAB*) had taught chemistry at Columbia College (New York) prior to his travels in Europe and, according to *BDAS*, he was "known as a teacher and disseminator of scientific discoveries from Europe." His most significant contacts were made in Paris where he attended numerous lectures, and spent considerable time with Gay-Lussac, Berzelius, Thenard, Cuvier, and Brogniart. In England he attended several of Sir Joseph Banks's *conversazione*, and met or heard lectures by numerous scientists, but he seems to have been more interested in English industry than science. He visited Birmingham, Manchester, Plymouth Dockyard, Sheffield, Leeds, and New Lanark, the last of which he described at length. In Italy, the Netherlands, Scotland, and Ireland Griscom largely visited learned institutions and hospitals. Griscom is perhaps the first American travel writer who acted as (and who was accepted as) an equal to the European scientists. As such he was able to provide the earliest comprehensive account of European science which was not merely a traveler's description, but rather a sophisticated assessment of individuals and institutions.

14.   *Harrison, Joseph, Jr., *The Iron Worker and King Solomon*. 2nd ed. Philadelphia: J. B. Lippincott Co., 1869. Russia and England.

Considerable sections of the idiosyncratic book relate to Harrison's years in Russia, 1843–52, where he built locomotives for the St. Petersburg-Moscow Railroad, and a bridge over the Neva River. He reminisced more on personalities in the Russian engineering bureaucracy than on the engineering works themselves. Harrison (1810–74, *DAB*) had been asked to go to Russia on the strength of his achievements in the Philadelphia firm of Eastwick & Harrison, and after his Russian experience he returned to pursue a successful career in that city. Overall this book's reminiscences are breezy and egotistical.

15.   Hazard, Erskine, "Observations upon Rail-roads," *Journal of the Franklin Institute* 3 (1827): 275–77.

Erskine Hazard, a Philadelphia civil engineer, was in England and Wales in 1826 to observe railroads, and returned to build the Mauch Chunk Railroad for the Lehigh Coal and Navigation Company. In this note he related some of his observations of British rails, roadbed, and carriages. He noted that steam locomotives had fallen out of favor on the Hetton railroad (to be replaced with stationary engines), and that he was "decidely of opinion, that in this country a rail-way of wood, sheathed with iron, would be preferable to any other, and could be kept constantly in order at the least expense." Hazard's Mauch Chunk Railroad was examined by many early American railroad promoters and engineers as the best example of railroad technology in America.

16.   *Henry, Joseph, *The Papers of Joseph Henry*. Nathan Reingold, *et al.*, eds. 4 vols. to date. Washington: Smithsonian Institution Press, 1972– . Numerous sketches. England, Scotland, France, Belgium.

Henry (1797–1878, *DAB*) was a student and teacher at Albany (NY) Academy before he moved on to Princeton College in 1832 and ultimately (in 1846) to head the Smithsonian Institution. Volume 3 of the *Papers* is dominated by Henry's trip abroad, March–October 1837, mainly to purchase scientific instruments for Princeton. That purpose reinforced Henry's own predilection for designing and making scientific apparatus, and makes his letters and diary entries read like a manual of experimental and teaching instruments at times. But he also reported on the major scientific institutions of the day, including the Royal Society, Geological Society of London, British Museum, Institute of France, Royal Society of Edinburgh, Royal Institution of Edinburgh, and the British Association for the Advancement of Science meeting.
    Henry recounted meetings with many British and French scientists, and a few Belgians. His publications obtained for him a positive reception by almost all with whom he had contact. Only a brief incident at the BAAS meeting in Liverpool pointed up a lingering superiority which some Europeans felt regarding American science. An impressive number of Americans are also noted in Henry's account, including several scientists and physicians, and many merchants. The editors also chose to include in this volume several entries from the diary of Alexander Dallas Bache, who accompanied Henry during much of his European visit.

17.   Hillard, George Stillman, *Six Months In Italy*. 2 vols. Boston: Tichnor, Reed, and Fields, 1853.

Hillard (1808–79, *Appleton's*), a Massachusetts lawyer, provided mostly standard tourist descriptions, but he did visit and describe the Arsenal at Venice, the University of Bologna, the Museum at Naples, and the Institution for the Insane at Rome. He was enthusiastic about the Museum of Natural History at Florence and was in particular awe of its wax anatomical models. He visited Italy in 1847–48.

18.   Hughes, George Wurtz, "Notes on Belgium," *Journal of the Franklin Institute* 3rd ser., 2 (1841): 73–83, 154–64, 224–32, 298–304.

Hughes (1806–70, *DAB*) was an army topographical engineer "sent to Europe to examine and report on public works" during 1840–41. This article is a series of dated entries written to the Corresponding Secretary of the National Institution in Washington, D.C. Hughes described the metallurgical industries of Belgium, including general statistics about mineral resources, and descriptions of the Royal Cannon Foundry at Liège, the great iron works of Couillet near Charleroi, and the nail works at Fontaine l'Eveque. He also examined the chemical and glass works at Oignies, and discussed at some length the building stones and cements of Belgium. Hughes had already been to England and Wales, according to this account, and subsequently went to the Netherlands.[1]

19. *Jefferson Thomas, *The Papers of Thomas Jefferson.* Julian P. Boyd and Charles T. Cullen, eds. 21 vols. to date. Princeton, N.J.: Princeton University Press, 1950– . Some sketches. England, France, Italy, Netherlands, Germany.

Volumes 8, 9, 11, and 13 contain some record of Jefferson's scientific and technical observations while the new republic's ambassador to France, 1785–89. His most substantial accounts were "Notes of a Tour into the Southern Parts of France. &c" (1787) in which he commented on agriculture and the Canal du Languedoc, and his "Notes of a Tour through Holland and the Rhine Valley" (1788) with observations on agriculture, viniculture, windows, canal machinery and other topics. Jefferson's agricultural interests are comparatively unusual among traveler's accounts, and perhaps thereby more important. It is notable, for example, that on his return from two months in England (a trip of which little record survives) he commented that "the mechanical arts in London are carried to a wonderful perfection," but what impressed him the most was a steam-powered grist mill (Jefferson to Page, 4 May 1786)—the industrialization of food processing. And while traveling through southern France he wrote that

In the great cities, I go to see what travellers think alone worthy of being seen; but I make a job of it, and generally gulp it down in a day. On the other hand, I am never satiated with rambling through the fields and farms, examining the culture and cultivators, with a degree of curiosity which makes some take me to be a fool, and others to be much wiser than I am. (Jefferson to Lafayette, 11 April 1787)

20. *Keating, William H., *Considerations Upon the Art of Mining. To which are added, Reflections on its actual state in Europe, and the Advantages which would result from an Introduction of this art into the United States. Read before the*

1. George Wurtz Hughes, *Report on Some of the most Important Hydraulic Works of Holland.* Washington: Printed by Order of Congress, 1843. (Not seen.)

American Philosophical Society, July 20th, 1821. Philadelphia: M. Carey and Sons, 1821. France, Switzerland, Savoy, Germany, Netherlands, Scotland, England.

Keating (1799–1840, *DAB*), was a Philadelphian who studied in Paris at the École des Mines, arriving in 1817. Upon his return he became an investor in enterprises developing the anthracite coal regions of Pennsylvania. In this substantial paper (87 pp.) Keating reviewed the state of mining technology in Europe and found it substantially in advance of American practice—in fact, he disdained even to call American ore extraction "mining," writing that "we may be warranted in saying that there are as yet no mines in activity in the United States." He concluded by arguing for the introduction of mining because of the economic benefit it would bring, and particularly urged the importation of German miners and steam engines.

Keating provided extensive descriptions of the mines of France, Germany, and Britain, and of the École des Mines.

21. Knight, Jonathan, William G. McNeill, and George W. Whistler, [Letters on British railroads], *Niles' Weekly Register* 34 (1829): 92–93, 273–74.

The Baltimore and Ohio Railroad commissioned these engineers to go to Britain to examine railroad practice before commencing construction of the B&O. The engineers never did publish a full report of their observations as they expected to, but three of their letters to the company were published. They discussed the Stockton and Darlington, and Killingworth railroads, and assessed the capabilities of current railroad technology, concluding that "experience daily develops the great advantages resulting from the introduction of rail roads."

22. *La Roche, René, "Medical Education and Institutions: An Account of the Origin, Progress, and Present State of the Medical School of Paris," *American Journal of Medical Sciences,* offprints for May 1831, pp. 1–16; August 1831, pp. 1–18; February 1832, pp. 35–72.

La Roche (1795–1872, *DAB*) was a Philadelphian who received his M.D. from the University of Pennsylvania in 1820, then went abroad to study, 1827–29. In this three-part article he wrote a history of the Medical School to 1830, followed by a lengthy description of the school's curriculum. La Roche was known as a "voluminous writer" who communicated to Americans much of what he learned abroad.[2]

23. *Livingston, Robert, *Essay on Sheep: Their Varieties—Account of the Merinoes of Spain, France, &c.* Concord, N.H.: Daniel Cooledge, 1813.

While Minister to France during the Jefferson administration, Livingston (1746–1812, *DAB*) acquired "such information in agriculture and the arts as would be useful to my fellow citizens." An especial interest of his was merino sheep, which he had sent to the United States in 1802. This book records his observations of European sheep-raising and -breeding practice, but it concentrates on good sheep husbandry in America.

24. *Merrick, Samuel V., *Report upon an Examination of Some of the Gas Manufactories in Great Britain, France and Belgium* . . . Philadelphia: Philadelphia City Councils, 1834. Scotland, England, France, Belgium.

In this 45-page pamphlet Merrick (1801–70, *DAB*) reported on the European trip he was commissioned to take by the Philadelphia City Councils. His comments are general and consider what is the best technical practice in Europe and what is most appropriate for Philadelphia. Investigating gas apparatus did not take all of Merrick's time: while he was in England he also obtained the American rights to the Nasmyth steam hammer.[3]

25. *Morgan, John, *The Journal of Dr. John Morgan of Philadelphia, from the City of Rome to the City of London, 1764.* Julia Morgan Harding, ed. Philadelphia: J. B. Lippincott Company, 1907. Italy, Switzerland.

During his English medical studies Morgan (1735–89, *DAB*) took a gentleman's tour of Italy, but gave considerable attention to scientific and technical sites. He met several Italian professors, including Dr. Laura Maria Catherina Bassi (Bologna), Dr. Serrati (Bologna), Dr. Morgagni (Padua), and Dr. Flaminio Torrigiani (Parma), and visited various anatomical collections and museums. At Venice he paid attention to glassmaking and the Arsenal's shops. Later he observed the waterworks of Geneva and spent considerable time with Voltaire.

2. S. v., "René La Roche," in *The Biographical Encyclopedia of Pennsylvania* (Philadelphia: Galaxy Publishing Co., 1874).
3. Bruce Sinclair, *Philadelphia's Philosopher Mechanics: A History of the Franklin Institute, 1824–1865* (Baltimore: The Johns Hopkins University Press, 1974), p. 321.

26. *Morris, Robert Hunter, "An American in London, 1735–36: The Diary of Robert Hunter Morris," Beverly McAnear, ed., *Pennsylvania Magazine of History and Biography* 64 (April 1940): 164–217, (July 1940): 356–406.

Morris (1713–64) was in England assisting his father, Lewis Morris, Sr., who was agent for a group of New York landholders. The two men had a variety of interests, including science and technology, which led them to visit the Chelsea Waterworks, a glasshouse, and the Royal dockyards at Chatham. They also had a scientific discussion with Lord Elmore and observed an attempt to induce rabies. Perhaps the most interesting detail in this diary is that the two Morrises were called upon to give their opinions about altering the waterwheels of the Faversham and Guilford gunpowder mills. Apparently Americans were already recognized for their innovative waterpower technology in the early eighteenth century.

27. Mott, Valentine, *Travels in Europe and the East.* New York: Harper & Brothers, 1842. Britain, Ireland, France, Belgium, Netherlands, Prussia, Saxony, Bohemia, Austria, Bavaria, Switzerland, Italy, Malta, Greece, Egypt, Turkey, Moldavia, Wallachia, Hungary.

The author (1785–1865, *DAB*) was a surgeon who was educated at Columbia College, then studied in England and Scotland in 1807. He traveled abroad again in 1834–41, renewing his acquaintances in Britain and making new ones elsewhere in the medical community of northern Europe. He sometimes combined accounts of his current travels with reminiscences of his school days. The book often makes assessments of the successes and failures of European surgery. He was enthusiastic about French accomplishments in orthopedics and the institution of charting in a Vienna hospital. He saw several medical and anatomical museums. Near Eastern and Balkan medicine were beneath his notice.

28. [Paine, Robert Treat], "Information to Students Visiting Europe," *American Journal of Science,* 2nd ser., 22 (November 1856): 146–48. England, France, Germany.

The author is identified as T. R. P., almost certainly an error for the initials of Robert Treat Paine (1835–1910, *BDAS*) who was abroad studying in England and France, 1856–57. Paine urged students who wished to obtain a liberal education in the sciences to come to Paris. He stated that there were British schools of quality in mineralogy, geology, and chemistry, but the cost of living in London was high. Germany he dismissed as having "schools celebrated for this or that specialty in science, but hardly any, where all [subjects] are taught by men of equal ability." (p. 147) Paine then discussed the various scientific institutions of Paris, remarking that "no where in the world can there be found as *clear* and *lucid* an exposition of the principles of all the sciences as at Paris." (p. 247) One might compare this with the experiences of Benjamin Smith Lyman (no. 60) or Alfred P. Rockwell (no. 70).

29. *Pancoast, Joseph, *Professional Glimpses Abroad: A Lecture, Introductory to a Course on Anatomy, in the Jefferson Medical College of Philadelphia, Delivered October 17, 1856.* Philadelphia: Joseph M. Wilson, 1856. England, France, Germany, Austria, Italy.

Pancoast (1805–82, *DAB*) was an anatomist and surgeon who taught at Jefferson Medical College. Having just returned from some months in Europe, where he observed surgical practice and visited medical institutions, Pancoast discussed his experience. He commented particularly on plastic surgery, bone and joint surgery, and ophthalmic surgery. He described at some length the use of chloroform and sulfuric ether (ethyl ether) as anesthetics; he was undecided as to which was better and what was the proper mode of administration. Pancoast's lecture is a clear example of how the latest news from Europe was disseminated.

30. *Peale, [Benjamin] Franklin, "Description of the new Coining Presses lately introduced into the U.S. Mint, Philadelphia," *Journal of the Franklin Institute,* new ser. 18 (November 1836): 307–10. France, Germany, England.

Peale (1795–1870) had recently studied mints in Europe (see no. 66) and, in describing the new steam coining press at the Philadelphia mint, he briefly mentioned his observations. He noted that the new press of his design was derived from those he saw at Karlsruhe and those of "Monsieur Thonnellier of Paris."

31. *Peale, [Benjamin] Franklin, "On the Manufacture of India Rubber Web," *Journal of the Franklin Institute,* new ser. 19 (February 1837): 109–12. France.

While in France in 1834 (see no. 66) Peale (1796–1870) had observed and taken notes on the making of elastic cords at St. Denis. In this article he described the ten steps of manufacture in some detail, with two slight sketches, although he admitted that he could not provide sufficient information for instituting manufacture.

32.　Peale, Rembrandt, *Notes on Italy*. Philadelphia: Carey & Hart, 1831. France, Italy, Switzerland, England.

This volume was published in journal form, recording some of Peale's thoughts and observations while abroad, 1828–30. Although he was an artist, Peale (1778–1860, *DAB*) demonstrated his family's catholic interests by visiting a variety of museums and industrial works. He was particularly impressed by the Jardin des Plantes at Paris, which he noted had enriched its collections noticeably since his visit of 20 years earlier, though he still thought its insects and animal specimens were inferior to those in the Philadelphia Museum of his brother Titian Peale.

33.　*Robinson, Moncure, "Letters of Moncure Robinson to his father, John Robinson, of Richmond, Va., Clerk of Henrico Court," *William and Mary Quarterly,* 2nd ser., 8 (April 1928): 71–95; 8 (July 1928): 141–56. France, Britain, Netherlands.

Robinson (1802–91, *DAB*) went to Europe in 1825 to study civil engineering and remained for two years, attending lectures in Paris and traveling in France, the Netherlands, England, Wales, and perhaps Scotland. Fifteen of his letters (mostly from France) were published without editing. I compared them with the originals in the Moncure Robinson Papers at the College of William and Mary and found them accurately transcribed.

　　The letters not only provide details about Robinson's education and travels, but frequently record his opinions about European technology. After a few weeks in France he went to London and did not hesitate to judge the French "at least one hundred years behind the English" in "practical mechanics."

34.　*Rush, Benjamin, *Letters of Benjamin Rush.* Lyman H. Butterfield, ed. Memoirs of the American Philosophical Society, vol. 30, pt. 1. 2 vols. Princeton, N.J.: Princeton University Press for the American Philosophical Society, 1951. Scotland and England.

About twenty letters in this collection date from Rush's medical studies in Britain, 1766–69. Rush (1745–1813, *DAB*) became deeply involved in the professional activities of both the students and physicians in Edinburgh and

London. He became Joseph Black's lecture assistant and claimed that Black "has honored me with his particular friendship." Although he found his Edinburgh studies to be of the highest quality, on going to London he found that he learned much by going to lectures and attending hospitals there.

As a professor at the University of Pennsylvania Rush taught a large number of the most prominent physicians and future professors of medicine in the early United States.

35. Schinz, Charles, "Extracts from the diary of Charles Schinz, Consulting Chemist of Camden, New Jersey, during a Journey in Europe," *Journal of the Franklin Institute,* 3rd ser., 32 (1856): 56–63, 129–37. Switzerland, Bavaria, Austria, England.

Schinz is unidentified except by the information in this article, which states that he was born in Switzerland and emigrated to the United States within the twelve years previous to publication. While in Switzerland he had been a consultant to a textile bleaching works. On this tour of 1855–56 he observed morocco tanning, concrete vaulting, chlorine bleaching of textiles, wood-pulp paper making, turf compressing for fuel, the making of illuminating gas from wood, and glass manufacture.

A considerable portion of this article is given over to Schinz's descrip-tion of the Polytechnic School of Switzerland. After listing the courses (which he notes follow the German plan) he states that "we may safely assert that this *is not* the best plan for the education of practical men!" On the basis of his experience in industrial chemistry he argued that the students were taught too much theory and not enough practice.

36. Sellers, George Escol, *Early Engineering Reminiscences (1815–1840) of George Escol Sellers.* Eugene S. Ferguson, ed. United States National Museum Bulletin no. 238. Washington: Smithsonian Institution, 1965. England. (Chapters 13–17).

Sellers (1808–99) was a mechanical engineer who was trained in his father's shop in Philadelphia. Like so many other American engineers of his time Sellers went abroad (in 1832) to study English technology, generally the most advanced in the world. A particular object of his attention was papermaking machinery and he contrived to visit the works of Bryan Donkin, the builder of the sophisticated Fourdrinier continuous-flow papermakers.

Although Sellers was impressed by the precision and careful construc-tion of Donkin's machines, overall he recognized that the American and English machines already embodied distinct traditions. No doubt reflecting in part a nationalist chauvinism, Sellers regarded English methods inferior in terms of specialization, types of tools, and scale of operations. (See no. 71.)

37. *Silliman, Benjamin, *A Journal of Travels in England, Holland and Scotland, and of Two Passages over the Atlantic, in the years 1805 and 1806.* 2 vols. New York: Ezra Sargeant, 1810. England, Scotland, and the Netherlands.

Benjamin Silliman (1779–1864, *DAB*) is the archetype of American reporters on European science and technology. Trained at Yale, he was asked to prepare to teach science there by studying further at the University of Pennsylvania. Then in 1804 he was commissioned by the Yale trustees to go to Europe to purchase books and scientific apparatus for the college, and to increase his own learning.

Silliman's journals of his travels was at least in part kept for his brother, Gold S. Silliman of Newport, Rhode Island, who was interested in textile manufacturing. But Silliman's descriptions range widely over sights, institutions, and individuals. He was an indefatigable visitor of museums, gardens (he bought a pass to the Chelsea Gardens in London), live-animal exhibits (he reported being spit upon by a llama) and private scientific cabinets. He saw numerous industrial and engineering sites, including cotton mills at Manchester (though he confessed to his brother that he could not possibly describe all of the processes), Liverpool and London docks, and the mines of Cornwall. He briefly visited the Netherlands and visited museums and collections there, but was denied permission to enter France.

Silliman found English scientists and manufacturers receptive to his inquiries and sympathetic to his purposes. Attending one of St. Joseph Banks's *conversaziones* he met numerous prominent scientists and subsequently met others at the Royal Institution, Cambridge University and Edinburgh University. If he did not move among them as a complete equal, there is no evidence that they lacked respect for his ability.

Silliman's journal of 1805–06 is the earliest major American travel account which I have found that focuses primarily on science and technology. It marks a divide between the dominance of accounts by the scientifically- and technically- informed who came from relative backwaters, and the beginning of travel by those who are purposefully reporting to a growing body of scientific and technical professionals in America. Silliman was himself conscious of America's changing role, and observed while at Cambridge University that "in classical learning and philological literature we are certainly far behind the English institutions, but in mathematics, ethics, and the physical sciences, some of our institutions are probably equal to them." (2: 233–34)

38.   *Silliman, Benjamin, *A Visit to Europe in 1851*. 2 vols. New York: G. P.
Putnam & Company, 1854. England, Wales, France, Sardinia, Tuscany,
Rome and the Papal States, Naples and Neapolitan states, Lombardy, Venice,
Switzerland, Germany, Prussia, Belgium.

Benjamin Silliman's second trip abroad followed his first by 45 years, a gap
which witnessed a transformation of American science and technology.
Silliman had played an important role in that transformation by founding the
American Journal of Science in 1818. Since his journal became a conduit for
the transatlantic flow of scientific ideas, Silliman was well known to Europeans by the time of his return visit. He moved easily among the famous
scientists of the era, including Charles Lyell, Adolphe Brongniart, Francois-
Jules Pictet, Justus Liebig, and Alexander von Humboldt. After a few days in
Paris Silliman could report that "among the men of science in that city we
met with only a solitary instance of cool manners."

During the six months he was abroad (as on his earlier trip) Silliman
frequently visited botanical and zoological gardens, and private collections of
specimens. Though he examined some chemical and physical apparatus, in
line with his own interests he focused on mineralogy, geology, paleontology,
and botany. He was perhaps most impressed by the "immense and unrivalled
treasures" of the Jardin des Plantes at Paris, where he reveled in the 200,000
specimens of extinct and extant animals. But even in the modest Sicilian city
of Catania, Silliman appreciated the mineralogical and geological collections
of its museum of natural history.

For all his scientific orientation, the ostensible purpose for Silliman's
visit was to see the Crystal Palace Exhibition in London, with its largely
technological exhibits. On his first visit late in March 1851 he was greatly
impressed by the building itself, but noted that many of the exhibits had yet
to arrive. Returning late in the summer he reported that he had "walked many
hours, and I presume ten miles in this immense structure, [yet] I seem only to
have begun to see it." He assessed the American exhibit as "not so splendid"
as the other exhibits, but noted that the plow and reaper were particularly
valued and appreciated for their utility.

Throughout his travels Silliman showed a keen interest in civil engineering. Between debarking at Liverpool and arriving at London a few days later
he saw the Liverpool and Manchester Railroad, the Liverpool docks, the
Menai Suspension Bridge, the Britannia (tubular) Bridge, the railway viaduct
and canal aqueduct at Chirk, and the Great Western Railroad. Later, on the
continent, he was impressed by French highways and bridges, but he noted
that "almost all Europe is superior to us in the United States" in that field.

Overall, this journal is more hectic and summary than that of his earlier
travels, but still provides a superb record of how an informed American

assessed the most important persons, institutions, and material culture of European science and technology.

39. *Silliman, Benjamin, Jr., "Miscellaneous Notes, from Europe," *American Journal of Science,* 2nd ser., 12 (September 1851): 256–61. Italy and France.

Benjamin Silliman, Jr. (1816–85, *DAB*) and his friend George Jarvis Brush (b. 1831, *Appleton's*) accompanied the elder Silliman on his European trip of 1851 (see no. 38). This article reports first on three natural phenomena of Italy—Mt. Vesuvius, the Grotto de Cane and Lake Agnano (near Naples), and the Sulphur Lake of the Campagna (near Tivoli), the last two renowned for giving off large quantities of volcanic gases. Following a brief note on the inactive meteorological station on Mt. Vesuvius, Silliman then provided a thorough description of Gillard's apparatus for making an intense hydrogen flame with the aid of a platinum catalyst. He observed it in use at the Christolef silver plate works in Paris.

40. *[Smith, George Washington], *Internal Improvement. Rail Roads, Canals, Bridges, &c.* Philadelphia: n.p., 1825.

George Washington Smith (1800–76) was an independently wealthy Philadelphian who offered this booklet (28 pp.) to the public at the height of the internal improvements fever in Pennsylvania. He claimed that it was in part "the result of personal observation, during a tour of Europe." Indeed, an obituary noted that he was "a frequent traveller and resident abroad," and the contemporary engineer, Moncure Robinson, himself a European traveler, thought highly of the booklet.[4] However, its comments on European technology are general, and Smith focused on promoting railroad technology in America rather than recounting his observations.

41. Stewart, F. Campbell, *The Hospitals and Surgeons of Paris.* New York: J. & H. G. Langley; Philadelphia: Carey & Hart, 1843.

This book was written especially for American medical students intending to study surgery in France, and was based upon the author's experience of several years residence there. Stewart argued that "in no other part of the

4. Horace Wemyss Smith, *Life and Correspondence of the Rev. William Smith, D. D.* 2 vols. (Philadelphia: S. A. George & Co., 1879), pp. 523–24; *Hazard's Register of Pennsylvania* 8 (December 1883): 362.

world can general or professional studies [in surgery] be pursued to greater advantage, or at so little cost to the student, as in France." Among the numerous advantages noted by Stewart were that anatomy could be more easily studied because of the abundance of corpses supplied by Parisian hospitals, and that orthopedics was practiced and taught as a separate branch of medicine. Much of the book provides descriptions of French medical institutions, and nearly half is devoted to biographical sketches of Paris surgeons.

42. *Strickland, William, "A Description of the Hetton Rail Road, in England; by Wm. Strickland, Esq. Civil Engineer. (With an Engraving)," *Journal of the Franklin Institute* 1 (1826): 15–16, plate. England.

Strickland's note is a matter-of-fact description of the construction and operation of the Hetton Railroad at Sunderland on the Wear River in northern England. During his visit in 1825 (see also no. 43) Strickland was impressed by the railroad's use of both stationary and locomotive steam engines to carry coal over a five-mile path with a series of steep inclines.

43. *Strickland, William, *Reports on Canals, Railways, Roads and Other Subjects, Made to "The Pennsylvania Society for the Promotion of Internal Improvement."* Philadelphia: H. C. Carey & I. Lea, 1826. Oversized plates. England, Wales.

This oversized volume with 51 pages of text and 71 plates is one of the great engineering documents of the industrial age. William Strickland (c. 1787–1854, *DAB*) was a trained and experienced engineer when he was commissioned early in 1825 by the Pennsylvania Society for Internal Improvement to travel to Britain to report on canals, railways, turnpikes, breakwaters, calico printing, gas lighting, machinery, and iron manufacture (topics of particular interest to Philadelphia merchants). During his six-month visit he was aided by "distinguished civil engineers" who "freely communicated to him the designs of their most important public works; [and] their plans of improvement, contemplated or commenced, were exhibited to him." In a series of subsequent reports, illustrated with excellent drawings by Samuel Kneass, who was his secretary and traveling companion, Strickland wrote concisely yet informatively about each mandated subject. Published in a limited number of copies by subscription, Strickland's *Reports* quickly attained the status of a classic on both sides of the Atlantic. It was particularly influential in its depiction of railroads as effective technological systems.

44.   Tallmadge, James, letters in *Journal of the American Institute* 1 (1835–36): 54–56, 157–59, 210–13, 177–79, 327–31, 385–88, 434–41, 486–88, 613–14; 2 (1836–37): 46–47, 107–11, 269–72. England, Wales, Scotland, Ireland, France, Italy, Russia, Netherlands, and Switzerland.

James Tallmadge (1778–1853, *DAB*) was president of the American Institute of New York, a society for the promotion of science and technology. While traveling abroad in 1835–36 Tallmadge wrote a series of letters (17 of which were published) regarding his observations. His letters typically described specific practices or innovations which he thought Americans should adopt, and also mentioned his visits to various institutions of learning. Although Tallmadge reported on a wide range of science and technology, he was most interested in textile manufactures, especially silk; agriculture and sheep-raising; civil engineering and architecture; and medicine.

45.   *Warden, David Bailie, *Microfilm Edition of the David Bailie Warden Papers*. Bayly Ellen Marks, ed. Baltimore, Md.: Maryland Historical Society, 1970.

David Bailie Warden (1772–1845, *DAB*), was an Irish expatriate who lived in New York and was confirmed an American citizen in 1804. He became a member of the staff of the American legation in Paris in the same year. He lost his position in 1814, but chose to remain in Paris and for the next 30 years served as a cultural intermediary for American visitors.

Warden cultivated French intellectuals by sharing with them his collection of American materials and by publishing several works on the United States. As a result he was elected to the Institut de France and various learned societies in Paris, in addition to becoming acquainted with a range of major figures in French science, engineering, government, and literature.

The Warden Papers contain letters of recommendation to Warden on behalf of American visitors, letters to Warden by American travelers, and some of Warden's outgoing correspondence. The list of those who contacted Warden is staggering, and includes many of those listed in this bibliography, such as Moncure Robinson (no. 33), Benjamin Franklin Peale (no. 31), James Tallmadge (no. 44), and Alexander Dallas Bache (no. 2). A sampling of other scientific, technical, and medical luminaries whose names I noticed in a rapid survey are Charles W. Storrow, William Crawford, John Locke, Thomas G. Clemson, and William P. C. Barton. Letters typically describe briefly the applicant's purpose in visiting Paris and ask for Warden's assistance in achieving it. A representative letter was written in September 1836 by Wistar Pennock of Philadelphia to introduce Dr. T. Franklin Hulme, a recent graduate of the University of Pennsylvania going to France "to perfect himself in Surgery, and in those branches of Chemistry connected with medical jurisprudence."

The testimony of letters by grateful Americans indicates that Warden was usually able to supply contacts or information which opened doors or permitted the newly arrived to adapt quickly. Increasingly, and especially in the 1830s, Warden was asked to provide American physicians and surgeons with placements in French hospitals and clinics.

While the Warden Papers contain only occasional accounts of American travelers' own European experiences, the collection is by far the best body of material which can be used to document who was a European visitor in the 1815–1844 period.

Original collections of Warden Papers are at the Maryland Historical Society and the Library of Congress. This microfilm is of the MHS holdings.

46. Watson, Elkanah, *Men and Times of the Revolution.* Winslow C. Watson, ed. 2nd ed. New York: Dana and Company, 1857. Reprint ed. by Crown Point Press, 1968. England, France, Belgium, Netherlands, and Germany.

Watson (1758–1842, *DAB*) spent 1779–84 abroad. He kept a journal of his travels which was considerably edited by a descendant before it was published. The printed version contains an account of a meeting with Franklin in 1779 during which Franklin turned Watson's attention to canals. Subsequently Watson traveled on canals in France (Canal du Midi), Belgium, the Netherlands, and England (the Duke of Bridgewater's Canal). His experiences are especially significant because of Watson's later role as a promoter of the Erie Canal.

47. Watson, Elkanah, *A Tour in Holland in MDCCLXXXIV by an American.* Worcester, Mass.: Isaiah Thomas, 1790.

Much of this book was taken wholly into the fuller account of Watson's European travels (no. 46). This is a series of dated letters (all from 1784) which often describe canal travel, notably on the Delft, Hague-Leyden, and Haarlem-Amsterdam canals.

48.   *Audubon, John J., Papers. 3 boxes. England.

There are 21 letters from Audubon (1785–1851, *DAB*) written during his British trips of 1827–28 (12), 1830 (3), and 1834–35 (5). These are uniformly disappointing in terms of describing the British scientific community. However, Audubon did make brief mentions of numerous institutions, individuals and patrons of science whom he visited in search of subscribers for his *Birds of America*.

49.   *Bache, Alexander Dallas, Journals, notes, and papers gathered in Europe. 1836–38. Stephen Girard Papers (microfilm), series II, reels 474–76. Sketches. Austria, England, France, Ireland, Italy, Prussia, Scotland, and Switzerland.

Although Bache published the results of his European travels on behalf of Girard College (no. 2), his original notes have additional information about the scientific and technical institutions he visited. I found, for example, 16 pages describing Bache's visit to Glasgow University and his attendance at lectures there, none of which was mentioned in the book.

The handwritten material consists of dated journal entries with Bache's personal comments, descriptions of schools, occasional sketches, and lengthy transcriptions of curricula, rules and regulations, and financial statements. The most remarkable manuscript segment is a diary of his visit to Paris, July–August 1837, in the company of Joseph Henry.[5] There is also a massive collection of books, pamphlets, and other printed documents containing annual reports, descriptions, and histories of the educational institutions Bache saw.

50.   Baldwin, Loammi, Jr., Diary of Travels. 1823. Sketches. England, France, the Netherlands (including Belgium).

Loammi Baldwin's personal journal is an excellent example of the gathering and recording technical information for future contemplation and use. Baldwin (1780–1838, *DAB*) was already a seasoned civil engineer by the time

5. Bache's European diary prior to his Parisian stay was partially published in Nathan Reingold, *et al.*, ed., *The Papers of Joseph Henry*. 4 vols. to date. (Washington: Smithsonian, 1972–   ), 3: 166–70, 206–09, 231–41, 243–46, 250–53, 297–98.

of this, his second trip abroad, and he was anxious to inspect hydraulic engineering sites. After brief notes on Dover harbor and the cotton mills and canal of St. Quentin (France), Baldwin's journal focuses on canals, roads, docks, and harbors in the Netherlands (then including Belgium). Nearly every technical description is accompanied by one or more careful pen-and-ink sketches. For example, Baldwin was impressed by the North Holland Canal and, in addition to lengthy comments, made sketches of lock gates, an inclined plane, sluice gates, excavating machinery, and a winch with man-wheels for lifting boats out of the water.

51. Fischer, Miers, Jr., letters to Thomas Gilpin, 1811–12 (copies), and journal of travels. Russia.

Two of Fischer's letters include his comments on industries at St. Petersburg. He was particularly fascinated by the manufacture and use of steam engines there, which he described as being made in "every force from a two horsepower up to 80 according as you order them; are entirely of cast iron beam and stand, and are as finely finished as a clock." His journal (also at the APS Library) has similar comments.

52. Girard, Stephen, Papers. 1769–c. 1840. Microfilm of original papers at Girard College.

Stephen Girard (1750–1831, *DAB*) was a wealthy Philadelphia merchant with an extensive overseas trade and in his later years substantial business interests. His massive collection of business records and manuscripts is available only on microfilm at the APS. Tables of contents and a subject-name index aid the researcher.

Letters from his American agents abroad and his ship captains and supercargoes sometimes comment on industrial matters. In a random search I noticed a letter from Edward George to Girard, dated from Liverpool on 24 September 1823, which reported George's conversations with Manchester cotton spinners regarding new textile mills in the vicinity. While other letters seemed to be confined to prices current and assessments of market movements, the correspondence covers so many products that it is a valuable index of Americans' knowledge of European technology.

53. *Hewson Family Papers. 1794–95. Microfilm. England.

Thomas Ticknor Hewson (1773–1848, *Appleton's*) was born in England, emigrated to the United States in 1786, and attended the University of

Pennsylvania medical school. He returned to Britain in the summer of 1794 to study at St. Bartholomew's Hospital in London, and the next year he went to Edinburgh for medical lectures there. He soon returned to practice in Philadelphia.

This collection has four letters written during Hewson's year in London. They describe his attendance at the hospital, which concentrated on dissections and anatomical studies, as well as the society of medical students. He also mentioned attending the Royal Society and meeting Sir Joseph Banks.

54. *Hutchinson, James, Papers. 1775–76. England.

Hutchinson (1752–93, *DAB*) was an American physician trained at the University of Pennsylvania.[6] Going to London in 1775 to study medicine, he was caught by the opening of the American Revolution, and after a year of studies had to find his way to France in order to get passage home.

Six of his letters record his course of study in some detail. Like so many of his American contemporaries, he relied upon Dr. John Fothergill for advice about how to carry out his studies. Following his recommendations, Hutchinson became a surgical dresser at St. Bartholomew's Hospital under Percival Potts, and attended anatomical and other lectures by William Hunter, John Hunter, Fordyce, and Potts. He quickly obtained the confidence of his mentors and within a few months was permitted to assist in or even carry out operations. Though Hutchinson complained of having little leisure time, he also noted that he had been elected to "one or two" of the local literary societies.

Near the conclusion of his studies Hutchinson assessed his American medical education as better than that available in London, but noted that the practical experience and education in surgery there far surpassed what he could have had in Philadelphia.

55. Kinloch, Francis, Letters and Papers. 1776–1809. Microfilm. Italy, France, England, Germany, Switzerland.

Francis Kinloch (1755–1826, *Appleton's*) was a South Carolinian who was educated at Eton, and subsequently took a European tour culminating in

6. Whitfield J. Bell, Jr., "James Hutchinson (1752–1793): Letters from an American Student in London," *Transactions and Studies of the College of Physicians of Philadelphia,* 4th series, 34 (July 1966): 20–25. Bell's article prints in part three of Hutchinson's letters from England.

residence at Geneva in 1775–77. These papers are largely letters and notes to Kinloch's friend, Johannes von Muller, a Swiss historian. They are tantalizingly brief on Kinloch's education under Charles Bonnet at Geneva, and rarely record his observations of European science and technology. His journal of travels in Switzerland, beginning 12 July 1776, mentions the harbor of Versoy and the machinery of the salt mine at Bex and is perhaps more revealing than his letters.

56. *Larrey, Baron Dominique-Jean, and Felix Hippolyte Larrey, Letters from Americans, c. 1818–1860. Microfilm.
Baron Dominque-Jean Larrey (1766–1842) and Felix Hippolyte Larrey (1808–95), father and son, were prominent French surgeons sympathetic to American students in Paris. These letters, dating from as early as 1828 and as late as 1860, are largely letters of introduction, letters of thanks for assistance, and correspondence. Daniel Brainard (1812–66, *DAB*), for example, studied in Paris in 1839–41 and 1853–54. On return from the latter trip he wrote to F. H. Larrey to thank him for friendship and assistance, and in 1858 he wrote to tell of significant surgery he had performed in Chicago.

57. *Lea, Issac, Journals. 1832, 1852–53. 16 volumes. Photocopies and microfilm. England, France, Belgium, Germany, Switzerland.

Lea (1792–1886, *DAB*) was a Philadelphia merchant who was a partner in the Carey publishing house from 1821 to 1851. He had decided interests in natural history, particularly geology, paleontology, and conchology, which formed the focus of trips to Europe in 1832 and 1852–53. His journals are detailed and often informative about individuals and collections. He did not hesitate to be critical of what he saw, in 1832 judging the British Museum's collection of minerals, shells, and birds very good, but the animals "decidedly bad."

Lea attended many professional gatherings. In 1832 he was at meetings of the BAAS and the Royal Society. He attended a session of the Institut de France and commented that "there must have been more than 100 savants present. Biot, Geof. St. Hillaire & some others were pointed out to me across the room. The meeting has a striking effect on a stranger." In 1852 he met for several days with the German scientific association.

Overall, the journals mention many prominent figures in Lea's fields of interest, many of whom took great interest in Lea's own researches.

58.   *Lesley, J. Peter, Papers. 1844–45. Originals and microfilm. England, France, Wales, Switzerland, and Germany.

These materials antedate Lesley's (1819–1903, *DAB*) development as a metallurgist and geologist, although he already had strong scientific and technical inclinations. The relevant letters (21) and journals (3) record his travels before and after a winter of theological studies at Halle in Germany. He visited several engineering works and made sketches of them, including the Canal du Midi, bridges over the Rhine, and the Strassbùrg railroad. He occasionally mentioned meetings with geologists, including Elie de Beaumont and Christian Leopold von Buch.

59.   Linnaean Society of London. Letters by and about Americans. 1738–1872. Microfilm.

The most valuable group of letters in these items selected from the Linnaean Society's collections is a group of ten letters in Latin written by Adam Kuhn to Linnaeus while Kuhn was studying in Upsala, London, and Edinburgh. There are other letters of introduction for American travelers and some correspondence from returned Americans.

60.   *Lyman, Benjamin Smith, Papers. Notebooks (box 43), 1859–62. France, Germany. Sketches.

Lyman (1835–1920, *DAB*) was a Harvard graduate employed by J. Peter Lesley to do a mining and metallurgical survey of Pennsylvania in 1856–57. After a brief stay with the Iowa State Geological Survey he went to Europe to study at the École des Mines at Paris (1859–61), and the Mining Academy at Freiberg (1861–62). There survive eight notebooks from those years, with notes of his travels and studies written in English, French, and German.
     The notes for France include some class notes on mine drainage, and considerable travel observations for southern France. He visited ironworks, mines, the arsenal of Toulon, and railroad works. His German lecture notes are more considerable and focus on geology and mining technology. The accompanying personal notes make it clear that he immediately sought out the company of other Americans studying at Freiberg and often spent time with them thereafter.
     Lyman subsequently worked as a consulting mining engineer in the United States, Canada, India and Japan.

61. *Maclure, William, Journals and Notes of Travels. 1805–13. 18 journals. Microfilm. Sketches.

Maclure (1763–1840, *BDAS*), an independently wealthy Scottish merchant, came to the United States in 1796 and ever afterward considered it his home. He conducted extensive geological researches during his travels in Europe and North America, and was a patron of scientists and scientific institutions.

   The journals are daily records of Maclure's observations of the landscape, including geology, agriculture, inhabitants, and sometimes industries. On occasion he remained in a city long enough to examine museums and personal collections or to meet resident scientists. Given Maclure's geological interests it is not surprising that mines and quarries of various sorts came in for considerable comment. He was especially fascinated by the Swedish mines with their unusual minerals. But Maclure took interest in nearly anything innovative or unique, and used four pages of his journal to sketch out a new mode of distillation for wine which he saw at Montpellier in France.

   Maclure's role in the American scientific community was so important that these journals can be taken as a significant measure of the body of knowledge available to it during his era.

62. *Mitchell, Maria, Papers. 1857–58. Microfilm, 9 reels. England, Scotland, France, Italy.

Mitchell (1818–89, *DAB*) was an astronomer who was appointed the first professor of astronomy at Vassar. In 1857–58 she toured European observatories and scientific institutions.

   The fragmentary material in the Mitchell Papers is difficult to make sense of: much of it seems to be reminiscences or journal entries copied years later. Updated and out-of-order journal entries also make the chronology difficult to reconstruct. Some of the longer passages include a description of a visit to Greenwich Observatory, a memoir of a conversation with Alexander von Humboldt, and a reminiscence of a visit with the Herschel family. One notable aspect of the collection is a large group of letters she received from others, which help to trace her travels.

63. *Morgan, John. Morgan-Dick Letters, 1763–65. 6 items. France, Italy, England. See no. 26 for Morgan's journal of his European travels.

John Morgan, Philadelphia physician, studied at Edinburgh, London, and

Paris, and traveled to Italy during his European stay. His letters to Alexander Dick in Edinburgh are full of information about his experiences.

Morgan may have been the first American physician to study in Paris, where he took a position under the surgeon to the Hôtel de la Charité. Morgan subsequently presented two memoirs to the Academy of Surgery at Paris and was elected a corresponding member.

64.   *Parke, Thomas, Journals. 1771–72. 3 vols. Microfilm. England and Scotland.

Parke (1749–1835) was a Philadelphia Quaker who studied medicine in London and Edinburgh. He kept a rather full journal which includes much more about his personal life and outside activities than it does about his studies. Still, he wrote about all of his mentors and professors and about the professional activities of the students. These aspects of Parke's experiences have been ably chronicled in an article by Whitfield J. Bell, Jr.[7]

Worth noting (and not fully covered by Bell) is Parke's spring 1772 trip from Edinburgh to London during which Parke visited the industrial cities of Glasgow, Sheffield, and Birmingham. On the trip he saw the Carron Iron Works, a ribbon factory, textile works, tool factories, Matthew Boulton's hardware, plate and button factory, and a lead mine. Later he visited a calico printing mill near London.

65.   *Patterson, Robert Maskell, Papers. 4 notebooks, 4 other items, 1810–11. France.

Patterson (1787–1854, *BDAS*), a Philadelphian, studied medicine and science in Paris and chemistry with Humphry Davy in London, 1809–23. He was later a professor at the University of Pennsylvania, the University of Virginia, and Director of the United States Mint in Philadelphia.

The four small notebooks in the collections contain notes on lectures at the Jardin des Plantes for 1810, including: Defontaine on botany, Tremery on physics, La Cépède and Dumeril on fishes, Lamarck on invertebrates, and other notes on trees and shrubs, and meteors. There are also separate notes on Thenard's and Gay-Lussac's experimental means of analyzing the constituents of plant matter.

7. Whitfield J. Bell, Jr., "Thomas Parke's Student Life in England and Scotland, 1771–1773," *Pennsylvania Magazine of History and Biography* 75 (1951): 237–59.

66. *Peale, Benjamin Franklin, Papers. 1833–35. Microfiche in *The Collected Papers of Charles Willson Peale and His Family*. Lillian B. Miller, ed. Millwood, N.Y.: Kraus Microform, 1980. England, France, Germany.

Peale was in Europe as an agent of the United States Mint (see nos. 30–31). In these eleven letters and a final report of 269 pages Peale described his visits to mints, mines, and refineries in England, France, and Germany. He valued the French mints because they were freely opened to him, thought that the mint at Carlsruhe was the "neatest and best arranged" of those he saw, and took a course in assaying in London. Although it might seem that minting is a rather narrow field of technology, Peale examined a wide range of allied industrial processes, including sulfuric acid manufacture, platinum refining, bronzing processes, balance-making, and waste-metal recovery. This collection is extremely useful for examining a wide range of European industrial technologies.

67. Peale, Rembrandt, Papers. 1802–03, 1808, 1828, 1829–30, 1832–33. Originals, and Microfiche in *The Collected Papers of Charles Willson Peale and His Family*. Lillian B. Miller, ed. Millwood, N.Y.: Kraus Microform, 1980. France, England.

Peale was abroad several times on artistic missions and on family business (no. 32). Occasionally in his letters written from abroad he commented on scientific or technical subjects. In 1808, for example, he reported excitedly that he had visited the museum of the Conservatoire des Arts et Métiers which he called "a glorious establishment, where you see models of all kinds of machinery, and in hundreds of instances the large machines themselves, even the largest ginnys and looms." In 1833 he wrote to his brother Titian about three different museum collections he had observed. Although there are 18 travel letters, they are generally weak on science and technology.

68. Peale, Rubens, Papers. 1802–03. Originals and microfiche in *The Collected Letters of Charles Willson Peale and His Family*. Lillian B. Miller, ed. Millwood, N.Y.: Kraus Microform, 1980. England.

Rubens Peale (1784–1865) wrote eleven letters while in England exhibiting a mastadon skeleton. Several of his letters comment on public and private collections in natural history, especially the Leverian Museum, the British Museum, and Pidcock's Menagerie.

69.   *Quetelet, Lambert A. J., Selected Correspondence. c. 1830–74. Microfilm.

Quetelet (1796–1874, *DSB*) was an astronomer and statistician who founded the Royal Observatory in Brussels. His work in statistics brought him international attention, and he subsequently carried on a considerable correspondence which he used to promote international cooperation in the sciences. This collection includes about 300 letters from Americans or APS members. Some letters are introductions for American travelers. I noted two letters by Matthew Fontaine Maury written in Europe subsequent to attending the International Maritime Meteorological Conference at Brussels in 1853, and three by Alexander Dallas Bache discussing his visits with French and English scientists in 1838 (see nos. 2 and 49).

70.   Rockwell, Alfred Perkins, Papers. 1857–59. 7 volumes and about 20 loose items. Sketches. England, Scotland, Germany and Belgium.

After an education at Yale and the Sheffield Scientific School, Rockwell (1834–1922) spent two years abroad studying mining. First he attended lectures given by John Percy at the Museum of Practical Geology in London, and then he enrolled at the Mining Academy in Freiberg, Saxony. Later he taught mining engineering in the United States.

   There are two volumes of notes on Percy's lectures (1857–58), two volumes of notes taken while traveling in Britain (1858), two volumes of notes from Freiberg (1858–59), and one volume on travels in Germany, Belgium, and Britain (1859). There are also loose sketches and copies of various documents from British collieries. The collection is exceptional in the details it contains on ore and coal mining technology, including pumping, winding, ventilating machinery, washing and ore separation equipment, and roasting and coking ovens. Rockwell also took notes on glassworks, ironworks, potteries, railroads, and building construction.

71.   Sellers, George Escol, Letterbook. 1832. 7 letters. England.

This letterbook contains letters Sellers wrote to his wife while sailing to England, and seven letters written during his first few days there, 17 September–13 October 1832. Sellers was abroad to investigate papermaking machinery (see no. 36), but, as these letters indicate, he took the opportunity to investigate other industrial technologies, including coal mining, iron manufacture, and machine shops.

72. *Smith, Thomas Peters, Diaries. 1800–02. 4 vols. Germany, Denmark, Sweden, France, Switzerland, England, Wales.

Smith (1777–1802) was a Philadelphian with decided chemical and mineralogical interests.[8] From 1800 to 1802 he traveled throughout Europe to further his interests, and died at sea on the return voyage to the United States.

    Smith's diaries are not complete for all of his travels (the circumstances of his lengthy stays in Paris are not recorded, for example), but for certain times and places, particularly Sweden and England, they are detailed and informative. Smith met and talked with a large number of prominent scientists, and technologists, presumably finding it easier to obtain audiences in his capacity as Secretary of the American Philosophical Society. Particularly notable are his acquaintances with J. A. H. Reimarus, J. G. Gahn, James Watt, Jr., William Reynolds, and Samuel Kier.

    During his time abroad one can detect a change in Smith's interests. He was at first a dedicated collector of minerals and admirer of others' collections. Then about the time he was in Paris, where he attended lectures (some notes are in one of the volumes), he took a more industrial orientation. By the time he arrived in Britain late in 1801 he was thoroughly fascinated by factories, mines, canals, and iron works. He took detailed notes of those industrial works and often accompanied them with sketches. Particularly interesting are his descriptions and sketches of glassmaking (including tinted drawings) and of railroads.

    This is the most comprehensive travel account of any that I examined at the Library. Smith's diaries are very similar to William Maclure's journals (no. 61) in their intensity, and should be compared with Maclure's regarding mineralogy and geology in southern France, Switzerland, Germany, and Sweden.

73. *Williams, Jonathan, Papers. 1771. Microfilm. England.

Jonathan Williams (1750–1815, *DAB*) was Benjamin Franklin's secretary while Franklin was in England representing various colonies. In May 1771 they were members of a party who traveled in northern England to take in scientific and technical sights. Williams recorded nine days of travel in 43 pages, giving ample room for comments. His lengthier descriptions include a marble-sawing mill, a trip on the Duke of Bridgewater's Canal, and a china

---

8. Wyndham Miles, "Thomas Peters Smith: A Typical Early American Chemist," *Journal of Chemical Education* 30 (1953): 184–85.

factory at Derby. They also visited Joseph Priestley, who entertained them with "a great many very pretty experiments in electricity," and Matthew Boulton's factory in Birmingham, which Williams admired for its variety of manufactures, including "metal buttons, all kinds of watch chains, plated silver, gilded metal, and gold and silver trays and utensils."

A summary of William's journal (of which there is another version at Yale) was published in *The Papers of Benjamin Franklin*.[9]

---

9. *The Papers of Benjamin Franklin*. Leonard Labaree and William B. Willcox, eds. 23 vols. to date. (New Haven and London: Yale University Press, 1959–   ), 18: 113–16.